The Politics of Oil
and Revolution in Iran

The Politics of Oil and Revolution in Iran

A Staff Paper by Shaul Bakhash

THE BROOKINGS INSTITUTION
Washington, D.C.

THE BROOKINGS INSTITUTION is an independent organization devoted to nonpartisan research, education, and publication in economics, government, foreign policy, and the social sciences generally. Its principal purposes are to aid in the development of sound public policies and to promote public understanding of issues of national importance.

The Institution was founded on December 8, 1927, to merge the activities of the Institute for Government Research, founded in 1916, the Institute of Economics, founded in 1922, and the Robert Brookings Graduate School of Economics and Government, founded in 1924.

The Board of Trustees is responsible for the general administration of the Institution, while the immediate direction of the policies, program, and staff is vested in the President, assisted by an advisory committee of the officers and staff. The by-laws of the Institution state: "It is the function of the Trustees to make possible the conduct of scientific research, and publication, under the most favorable conditions, and to safeguard the independence of the research staff in the pursuit of their studies and in the publication of the results of such studies. It is not a part of their function to determine, control, or influence the conduct of particular investigations or the conclusions reached."

The President bears final responsibility for the decision to publish a manuscript as a Brookings book. In reaching his judgment on the competence, accuracy, and objectivity of each study, the President is advised by the director of the appropriate research program and weighs the views of a panel of expert outside readers who report to him in confidence on the quality of the work. Publication of a work signifies that it is deemed a competent treatment worthy of public consideration but does not imply endorsement of conclusions or recommendations.

The Institution maintains its position of neutrality on issues of public policy in order to safeguard the intellectual freedom of the staff. Hence interpretations or conclusions in Brookings publications should be understood to be solely those of the authors and should not be attributed to the Institution, to its trustees, officers, or other staff members, or to the organizations that support its research.

Foreword

SINCE February 1979 it has been difficult to ignore the effect of the Iranian revolution on the international oil market. A little more than a year after the revolution, oil prices had almost tripled, largely because of the disruption of Iranian supplies and the decision by the authorities of the Islamic Republic to reduce Iranian crude-oil exports significantly. In September 1980 Iran and Iraq became engaged in a war. The conflict further curtailed Iranian oil exports, and in Iraq daily production fell from 3.5 million barrels to well under 1 million barrels. After nearly two years during which Iranian production remained low, Iran began to market its oil aggressively in the summer of 1982, undercutting the prices and substantially exceeding the production ceiling set by the Organization of Petroleum Exporting Countries.

These events demonstrate the continuing importance of Iran as a Persian Gulf oil producer and the need to understand the determinants of its oil policy. In this paper, Shaul Bakhash examines elements of continuity and change in Iran's post-revolution oil strategy, the domestic forces that helped to shape oil policy, the regional repercussions of the Iranian revolution, and the manner in which the Iranian authorities have reacted to changes in the oil market over the past three and a half years.

Shaul Bakhash is currently a visiting professor in the Department of Near Eastern Studies at Princeton University. He is grateful to John D. Steinbruner and William B. Quandt for their helpful suggestions; to James Hitselberger for his research assistance; to Caroline Lalire for editing the manuscript; to Clifford A. Wright for checking its factual content; and to Ruth E. Conrad for patiently typing its many versions.

vii

This study is part of a research program on energy and national security supported by the U.S. Department of Energy. The views are those of the author and should not be ascribed to the Department of Energy, to the persons whose assistance is acknowledged above, or to the trustees, officers, or other staff members of the Brookings Institution.

BRUCE K. MAC LAURY
President

September 1982
Washington, D.C.

The Politics of Oil
and Revolution in Iran

IRANIAN oil production dropped precipitously as a result of the Iranian revolution of 1979. Iran was producing over 5.5 million barrels of crude a day in mid-1978. By late 1981 daily production was running at barely 1.2 million barrels. This dramatic drop was not primarily the result of falling reserves, insurmountable technical problems, or even simple economic calculation. Rather, it was rooted in revolutionary ideology, in domestic turmoil and the dislocations generated by the Iran-Iraq war, and in the government's inability to adjust policy to increasingly unfavorable market conditions.

It is possible, of course, to discern continuities in Iranian oil policy despite the crisis of the revolution. But post-revolution Iran poses problems of a different order from those prevailing under the shah. New men have been in charge of formulating and implementing oil policy. Often they start from different premises and are actuated by different economic and political goals.

The formulation of oil policy has been subject to powerful political crosscurrents that were not an important consideration four years ago. Domestic turmoil has spilled over into the oil industry, affecting operations. Different political factions struggling for power in Iran used the emotional issue of oil to advance their ends. Purges in the oil industry and hostility to foreigners undermined the country's technical and production capacity.

The revolution has created conditions that sharpen the sources of friction between Iran and the Arab states of the Persian Gulf. This has affected the thrust of Iran's policies inside the Organization of Petroleum Exporting Countries (OPEC). In the case of the Iran-Iraq war, which is

1

a direct consequence of the Iranian revolution, regional problems have resulted in considerable damage to both the Iranian and the Iraqi oil industries.

Nonetheless, Iran remains heavily dependent on oil, which continues to account for over 80 percent of all government revenues and now constitutes over 90 percent of foreign exchange earnings. Despite a general official commitment to keep oil exports down, revenue requirements are likely to exert pressure for higher levels of production and exports.

Oil policy, then, has had to be formulated under volatile conditions and often to satisfy the demands of conflicting policies and goals. At the same time, oil policy has had to be adjusted to changing market conditions: initially to a market in which demand was high and prices strong, and subsequently to a market in which demand and prices were rapidly falling. This paper examines the various ideological, political, regional, and economic factors that have helped to shape oil policy in post-revolution Iran and the revolutionary government's attempts to respond to changing market conditions.

Ideological Factors

The men who seized power in Iran in February 1979 came to office with ideological baggage, and a clutch of economic ideas, that had important implications for Iranian oil policy. They believed that the shah had been maintaining oil production at levels far beyond Iran's revenue requirements, that oil income was being squandered and the country's chief natural resource was being needlessly exhausted. They were suspicious of the international oil companies. They thought that oil exports should be limited, that the oil itself should be used in Iran to promote petrochemical and other oil-related industries, and that the operations of the large international oil companies (the "majors") in Iran should be ended or at least sharply curtailed.

These attitudes were not necessarily confined to the shah's active opposition. They came to be shared by economists and technocrats inside the Pahlavi government.[1] On the eve of the revolution there was

1. See, for example, the views expressed by Dr. Fereioun Fesharaki, *Revolution and Energy Policy in Iran,* EIU Special Report 82 (London: Economist Intelligence Unit, 1980).

already a growing consensus within the administration in favor of lower rates of oil production and more restrained levels of development spending. Insofar as the oil companies were concerned, the shah himself had effectively used opportunities to raise oil prices and whittle down the hold of the oil companies over the Iranian oil industry. The new agreement negotiated with the Consortium (the eight major and several smaller international oil companies) in 1973 considerably reduced the role of the majors in Iran. On the eve of the revolution in 1979, negotiations for a further revision of this agreement were under way.

Development of Radical Opinion

During the 1970s groups opposed to the shah developed more radical views on oil policy. Iran's oil resources, it was argued, were not only being wasted; they were also being plundered. The oil companies were not merely seeking to maximize profits. Along with their governments, they were party to a conspiracy to exhaust Iran's mineral resources, to undermine native Iranian industry, and to make the country dependent on the West and a consumer of Western products. The shah, in pursuit of his grandiose military and economic programs, was not simply an unwitting tool of the United States and the Western European countries; he was a willing lackey, an active partner, in the despoilation of Iran.

Such ideas were not new. They stemmed from the experience of real or perceived exploitation of Iran by the great powers during the nineteenth and twentieth centuries. For many Iranians, the perfidy of the oil companies, the United States, and the shah had been confirmed by the overthrow of Mohammad Mossadegh in 1953 and by the subsequent failure of the nationalist movement to realize its aspiration toward an oil industry fully independent of the oil companies and their governments. The shah, in fact, was never able to live down his role in the overthrow of Mossadegh and in the "sell-out" of the nationalist cause, and his collaboration with the United States to achieve those ends. Despite his considerable achievements in the 1970s in raising oil prices and in dealing with the international majors, the shah's oil policies continued to be suspect in the eyes of nationalist and opposition groups.

There was thus a historical legacy of mistrust of the international oil companies and skepticism of the ability, or willingness, of the shah and his government to defend Iranian interests against them. But in the 1970s these ideas came to be articulated with a sharper edge and gained in apparent coherence. This was due in part to the growing hostility to all aspects of the shah's government and in part to the opposition's criticism of oil policy, which was now being expressed in the context of a world economic system in which the European powers, the United States, and the multinational companies were, by definition, exploiters of the developing, or third world, countries.

It was part of the common stock of ideas and political theories of such left-wing opposition groups as the Mujahedin-e Khalq and the Cherik-ha-ye Fada'i-ye Khalq that Iran was being exploited by the industrial powers, the monarchy, and a cooperative ruling class. Ayatollah Khomeini routinely referred to the plunder of Iran's oil resources by the United States and the West (and of her gas resources by the Soviet Union) in his sermons and declarations during his exile in Iraq and Paris. "They [the shah and his officials] have committed treason against us," he noted. "They are giving our assets to America. They are giving our oil to America. They are giving it away at such a rate that in 30 years, we will have no oil left . . . and the nation gets nothing in return." A favorite theme was that the United States was giving Iran "scrap iron," or useless arms, in exchange for valuable oil. "They take away the oil gratis," he said. "They take it away and in return build military bases for themselves. They give us arms . . . which are intended for American use."[2]

Bani-Sadr's Views

Abol-Hassan Bani-Sadr, a close aide and adviser to Khomeini before the revolution, and a minister of finance and president of the Islamic Republic after the overthrow of the shah, developed the fullest and most systematic exposition of this particular perspective on the workings of the oil industry.

Bani-Sadr was the son of an Islamic cleric. As a student at Tehran University in the late 1950s and early 1960s, he had been active in the pro-Mossadegh National Front movement. He saw himself as an heir to the Mossadegh tradition and regarded the goal of freeing the Iranian oil industry from Western domination as the central pillar of that heritage. In the early 1960s Bani-Sadr settled in Paris, where he came under the

2. Ruhollah Khomeini, *Neda-ye Haqq* (collection of speeches and interviews published by the Federation of Iranian Students' Islamic Societies in Europe and America, 1979), pp. 206, 219.

influence of writers who were developing Marxist and left-wing critiques of Western and particularly American capitalism and imperialism. The French sociologist Paul Vieille was a mentor and colleague.

Oil remained a central concern in Bani-Sadr's work. With Vieille, he edited a collection of essays on oil under the title *Pétrole et violence*. He himself devoted a whole book in Persian to the subject, entitled *Naft va Salteh* ("Oil and Domination").[3] In these books Bani-Sadr expounded more fully the concept of oil as an instrument for the exploitation and domination of the developing and oil-producing countries by the industrial powers.

With regard to Iran, Bani-Sadr argued that economic policy itself, expressed in a series of five-year development plans, had been designed to turn Iran into a "dependent" country, condemned to export raw materials and to rely on the United States and Western Europe for manufactured goods. The five-year plans were in fact consciously designed to further the interests of the industrial powers. No real Iranian industry had been allowed to develop. Assembly plants were a sham. Iranian agriculture was being destroyed. The development of ports, the construction of highways from the ports to the interior, and the expansion of the railway network were all designed to facilitate the flow of a flood of Western goods into the Iranian market.

For Bani-Sadr, oil was the pivot on which this entire edifice revolved. It was critical to the West as a source of energy, raw materials, and vast profits. It provided the revenues with which Iran could pay for Western products. It served as an instrument for the domination of the Iranian economy, currency, banking system, trade, industry and, eventually, Iran's army, culture, and arts.[4]

While the industrialized states used oil to develop their economies and expand their industrial base, oil revenues, Bani-Sadr believed, were used by Iran in such a way as to destroy the foundations of the economy. "Instead of an economy," he wrote, "[oil] creates a 'sucking machine' that increasingly and more extensively over time soaks up the oil, other resources and the fruits of the labor of the people, and exports these to

3. *Pétrole et violence* was published by Editions Anthropos, Paris, 1974. Bani-Sadr's views are summarized from Abol-Hassan Banisadr, *Naft va Salteh* ([Tehran], 1977). For a less strident summary of these views, see Abol-Hassan Banisadr and Paul Vieille, "Iran and the Multinationals," and "The Present Economic System Spells Ruin for the Future: An Interview with Abol-Hassan Banisadr," in Ali-Reza Nobari, ed., *Iran Erupts* (Stanford: Iran-America Documentation Group, 1978), pp. 24–33, 117–39.

4. Banisadr, *Naft va Salteh*, pp. 331–34.

the industrial states. The gap between the owners of the resource and its real users grows regularly wider. What remains is a bitterness in whose flames are consumed the children and the capabilities of the oil-producing nation."[5]

Bani-Sadr described the relationship between oil company and producing state as one of "plunder," Iranian agreements with the oil companies as "imposed," and the extension of these agreements by the shah as "treason." He produced tables of statistics to show that OPEC price increases had failed to maintain pace with the deterioration in the terms of trade of the producing countries. Because oil at the petrol pump sold at twenty times the price it fetched at the wellhead, he reasoned that for every $20 billion in oil revenues, Iran was actually losing $400 billion.[6]

Between 1960 and 1970, he argued, OPEC's achievements had been insignificant. The dramatic price increases recorded after 1973, on the other hand, were engineered by the United States to save a falling dollar by forcing up the price of American oil, to wreck the economies of the European states, which could not afford high oil prices, and to create markets for American goods in the OPEC states.[7] OPEC, in fact, "has always been an instrument of the United States," and Iran and Saudi Arabia acted inside OPEC as the agents of American interests.[8]

Such a world view would imply significant changes in Iranian oil policy and in relations with the international oil companies. However, no sudden change occurred in oil policy in post-revolution Iran. The ideas held by Bani-Sadr and like-minded Iranians continued to subsist side by side with attitudes that, though critical of the shah's policies, did not imply so radical a transformation of past practice.

Radicalization of Oil Policy

One of the first acts of the new revolutionary government was to abrogate unilaterally the agreement under which the Iranian Oil Participants (the former Consortium) acted as purchasers of the bulk of Iranian oil and provided a broad range of services through their jointly owned Oil Service Company of Iran (OSCO). Announcing the abrogation in a speech to employees of the National Iranian Oil Company on February

5. Ibid., p. 21.
6. Ibid., pp. 51–54.
7. Ibid., pp. 27–29.
8. Interview with *Al-Nahar Arab Report and Memo* (Beirut), January 29, 1979.

28, 1979, the new NIOC chairman and managing director, Hassan Nazih, said, "We tell those companies that were imposed on us in the past that it is better for them to withdraw, because if they refuse, the workers will kick them out."[9]

Yet this dramatic measure was not as radical as it might appear. The shah's government, as noted, was already in the process of renegotiating the agreement with the oil firms when the revolution occurred. By 1978 many of the terms of the agreement were in any case a dead letter. The Bakhtiar government, which held office in the brief period between the shah's departure from the country on January 16, 1979, and the overthrow of the monarchy five weeks later, had drawn up a plan to reduce greatly the role of the former Consortium companies in the oil industry, with the NIOC directly hiring the technicians then working in Iran under contract to OSCO. In fact, the first post-revolution government and the administration at the NIOC were generally moderate in their oil policies.

NIOC and government officials were uniformly eager to get oil production and exports started up again, and the first tanker was loaded at the island of Kharg as early as March 5. The Bakhtiar government had banned exports to South Africa and Israel, but the new administration was not in a hurry to add others to the list or to boycott the major industrial countries. (Morocco and the Philippines were blacklisted, but these were not important customers of Iranian oil.) The chairman of the NIOC, Hassan Nazih, was a prominent civil rights lawyer and a man of the political center. At a time when sweeping purges were taking place in other government agencies, the "purge" in the NIOC under Nazih was relatively mild; he made clear he opposed policies that would force Iranian experts who were needed in the oil industry to leave the company.

Feelings against foreign staff were running high, and Nazih took the position that the NIOC could manage very well with only 10 percent of the foreigners who had been working in the oil industry before the revolution. But he did not rule out the return of some of these foreigners. By August one of Nazih's chief aides was asserting that foreign technicians would be required for highly technical work, such as gas reinjection and specialized processing. Nazih announced Iran's intention to play a vigorous and more militant role in OPEC; however, he could hardly have said less under the circumstances.[10] And although both inside and

9. *Middle East Economic Survey* (Nicosia), vol. 22 (March 5, 1979), p. 8 (hereafter *MEES*).

10. Interview with Mohammad Ali Movahed, *MEES*, vol. 22 (August 27, 1979), p. 4; and interview with Nazih, *MEES*, vol. 22 (March 19, 1979), p. 9.

outside OPEC, Iran took a generally hawkish stand on prices, that was
not much out of line with previous Iranian practice.

Limitation of Foreign Involvement

There were, however, several important developments. First, the
government acted to reduce foreign involvement in the oil industry.
After taking over the functions of the Oil Service Company, Iran in the
summer of 1980 also took over the operations of the NIOC's four foreign-
operated joint ventures and placed them under the control of the newly
formed and wholly owned entity, the Continental Shelf Oil Company.
In September the NIOC established a subsidiary, Kala, which in the
next year took over purchasing functions for the oil industry from the
Iranian Oil Services, the London-based subsidiary of the Iranian Oil
Participants.

Second, unlike the shah's government, the Islamic Republic displayed
little interest in downstream operations or in oil-related enterprises
outside Iran. The tendency was to withdraw from such activity. In April
1980 Iran notified its South African and French partners of its decision
to divest itself of its 17.5 percent share in a joint-venture oil refinery in
Sasolburg, South Africa. In July Iran sold to its South Korean partner
its 50 percent share in the Korea-Iran Petroleum Company refinery in
South Korea. Each of these joint ventures entailed an undertaking by
Iran to supply crude oil to the refinery.

Third, decisions were taken that affected the broad lines of energy
policy. The shah's ambitious nuclear energy program was virtually
abandoned. When the revolution occurred, work was already at an
advanced stage on two nuclear reactors being constructed at Bushehr
by West Germany's Kraftwerk Union. Work on these reactors has not
been resumed, and an initial, though unofficial, commitment to complete
them appears to have been abandoned. Work was also stopped on two
further nuclear reactors near Ahwaz, contracted to France's Framatome.
Construction was only at the initial stages, and the project appears to
have been scrapped.

The government also canceled plans for the construction of a second
Iran gas trunkline (IGAT II) to the Iran-Soviet frontier. In addition to
supplies for domestic consumption, the pipeline was slated to deliver
1,300 million cubic feet a day of nonassociated natural gas from the
Kangan and Nar fields in southern Iran, 1,350 kilometers from the Soviet

border. The bulk of the gas, in a swap arrangement, was intended for West Germany, France, and Austria.

Already before the revolution, studies undertaken for the government by Iranian analysts had questioned on economic and technical grounds the size of the intended nuclear power program and the viability of the second gas trunkline. The decision to abandon these projects under the Islamic Republic owed something to the earlier studies and recommendations. It also reflected the general tendency of officials in the Islamic Republic to avoid long-term, expensive commitments in the energy field and their inability, in post-revolution conditions, to carry out complicated negotiations and to make decisions involving large and controversial investments.

Iran before the revolution was also delivering 700 to 800 million cubic feet a day of associated natural gas from the southern oil fields to the Soviet Union through the existing gas trunkline (IGAT I). Gas deliveries were interrupted during the strikes that swept the oil fields in the fall of 1978, on the eve of the revolution. Deliveries were resumed, though in smaller quantities, after the revolution, and interrupted again in the spring of 1980 in a dispute over prices, which Iran considered too low. Gas deliveries through IGAT I were not, as of the summer of 1982, started up again.[11] The price dispute remained unsolved. The fall in oil production had reduced the amount of associated gas available, and the government wanted to retain the gas for domestic use, as a substitute for kerosene and other fuel oils, especially in view of the high cost of installing new refinery capacity.

New Marketing Patterns

Finally, changes occurred in crude marketing patterns, a development that was to have important medium-term repercussions on Iran's ability to market her crude during the oil glut that began in the second half of 1981 and intensified toward the end of the year.

The changes that occurred in marketing patterns were the result of both fortuitous market conditions and deliberate policy. The participating oil companies eventually signed purchasing contracts with the Islamic Republic, leaving for later negotiation their claims to compensation for

11. A more detailed account of Iran's pre- and post-revolution gas and nuclear energy programs is given in Bijan Mossavar-Rahmani, *Energy Policy in Iran: Domestic Choices and International Implications* (Pergamon, 1981), pp. 70–84, 105–21.

losses suffered because of the canceled agreement between Iran and the former Consortium companies. But the majors were initially reluctant, or at least cautious, purchasers of Iranian oil. The NIOC could not in any case have reverted to pre-revolution marketing patterns, which the revolutionaries had denounced as exploitative, and marketed the bulk of Iranian oil through the international majors.

Moreover, in a market short of crude oil, there were numerous ready customers for Iranian oil: independents, companies like the Japanese ones that in the past had purchased significant quantities of Iranian oil through the agency of the international majors, and governments in short supply. Furthermore, these buyers were ready to pay premium prices on shorter-term contracts, and they had appeared at a time when the new administration at the NIOC was trying to find its feet, without making long-term commitments, and was eager to retain flexibility until basic policies could be determined.

A pattern was thus established under which Iran sold oil directly to a wide variety of customers. By early May 1979 the NIOC was selling 2 to 3 million barrels a day (mmbd) under thirty-five contracts to the eight majors, twenty-one independents, and six governments. Within weeks the number of customers had risen to over fifty. A few of the sales contracts were for substantial volumes; for example, British Petroleum was initially lifting over 350,000 barrels a day and the Royal Dutch/Shell Group 200,000. But the majority of contracts were for small amounts of 20,000 to 50,000 barrels a day.

There was of course some precedent for direct NIOC sales. The NIOC had dealt outside the Consortium framework with companies and governments involved in offshore concessions. It was beginning to market directly larger quantities of its own oil and was making oil available to governments on a sale, barter, or joint-venture basis. This trend accelerated after 1976, when in view of falling demand the majors refused to lift their "nominated" volumes of crude.

Nevertheless, the pattern that emerged after the revolution, with Iran's large number of customers buying relatively smaller quantities on short-term contracts, was clearly new. NIOC officials were not entirely happy with the situation. In the months following the establishment of the Islamic Republic they repeatedly stressed their preference for dealing with "people we know and are accustomed to,"[12] and suggested on

12. *MEES*, vol. 22 (August 27, 1979), p. 2.

many occasions that there would be an early return to more traditional marketing practices. They recognized the advantages of dealing with the majors and bulk buyers with access to sophisticated downstream operations. However, for a number of reasons, the new pattern persisted.

NIOC officials remained hesitant about dealing in a big way with the former Consortium companies, which remained suspect in the eyes of both hard-line clerics and left-wing groups. The NIOC, in fact, had to justify continued sales to the majors in the face of widespread criticism. The desire to maximize per barrel revenues encouraged sales on short-term contracts to smaller customers. Political developments also continued to militate against sales to the majors.

The seizure of the American embassy in Tehran in November 1979 led President Jimmy Carter to ban oil imports from Iran and the Iranian government to ban oil sales to American companies. Trade sanctions, imposed by the United States in April 1980 and by the European Community in May, brought about changes in trade patterns that led to the sale of more oil to countries like Turkey and the Soviet bloc states, sometimes on a barter-type basis. Uncertainty in the Iranian government about long-term oil policy as well as general turmoil favored the retention of some flexibility—another reason for choosing short-term contracts and diversified customers.

The revolutionary government also made it a regular practice to sell directly on the spot market. Spot sales were intended to "test" the market. In the spring of 1979 Iran was returning to the market after a stoppage of exports that had lasted several months, and the new men heading the NIOC were uncertain about market conditions. Initially, large quantities were sold on the spot market. Officials downplayed such sales and repeatedly said that spot sales would not become a regular feature in Iranian sales strategy. But selling on the spot market also became a regular feature of post-revolution marketing policy, at least as long as spot prices remained highly attractive.

It became official policy to dispose of between 5 and 10 percent of Iranian crude oil through spot sales. Although the official reason was the desire to test the market, the high prices prevailing on the spot market, and the large amounts of uncommitted crude available to NIOC, clearly exerted a strong attraction. Indeed, there was a growing temptation to regard spot prices as the real markers for determining official prices. In late 1979 the NIOC attempted to sell crude to its regular customers at spot prices, and when negotiating new contracts for 1980, it sought to impose on these buyers a formula under which the prevailing spot price would influence price calculations.[13]

Under the impact of these developments, the pattern of Iran's oil sales underwent a change. Before the revolution the major international oil companies had lifted some 90 percent of Iran's crude exports under long-term agreements. After the revolution the majors' share of Iran's crude declined. Nine-to-twelve-month contracts, a wider variety of customers, and a tendency to bargain for short-term gain became more characteristic of marketing practice. Iran gained a certain reputation for unreliability as a supplier, and its relations with large customers were disrupted. This became a factor of some importance when world demand began to fall and prices to weaken in late 1981 and early 1982.

In the months after the revolution, moreover, virtually all groups both inside and outside the government were in favor of lower levels of production and exports. The government announced it would reduce production from a pre-revolution level of over 5.5 mmbd to about 3.5 to 4.0 mmbd. This was a sharp reduction but not an unreasonable one in the circumstances. Per barrel oil revenues were higher than before the revolution because of rising prices. And revenue needs were considerably lower because of government plans to drastically cut projected arms purchases and to eliminate large and expensive military and civilian construction programs and white-elephant industrial projects, including the multibillion-dollar nuclear reactor program.

Given the turmoil in the country, oil production remained surprisingly steady, at the planned levels, throughout 1979. Official production policy remained almost unchanged even after Nazih was dismissed as NIOC chairman in September. In March 1980, at the end of the first full year of oil operations under the revolutionary government, average daily production stood at 3.4 million barrels for the preceding twelve-month period. The figure would have been higher but for a slow start-up in the early weeks and a decline in production in the closing weeks of the Iranian year (which runs from March 21 to March 20).

Yet despite announced policy, production began to grow sluggish in early 1980 and continued to drop in the ensuing months. By September 1980, when the outbreak of the Iran-Iraq war disrupted Iranian oil exports altogether, Iran was exporting under 700,000 barrels a day. This change in intended export levels was a product of the political, ideological, and economic forces acting on oil policy in revolutionary Iran.

13. Ibid., vol. 23 (November 12, 1979), pp. 1–2, and vol. 23 (December 17, 1979), pp. 4–5.

Political Factors

Domestic politics and the struggle for power among the warring factions in the Islamic Republic were one factor that influenced the drop in oil production. Nazih was dismissed as NIOC chairman in September 1979, ostensibly for incompetence and mismanagement of the oil company. But the cause for his dismissal lay elsewhere. In May, in an address before a lawyers' association, Nazih had described Khomeini's goal of establishing a government based exclusively on Islamic law as "neither practicable nor desirable." He had criticized the activities of the revolutionary courts and policies that were driving well-educated Iranian technocrats out of government and out of the country. He had decried what he described as the "economics of the beggar bowl"—the inclination to regard the Iran that the revolutionaries had inherited from the previous regime as an economic ruin and to emphasize the poverty of the people rather than the opportunities oil revenues offered for rapid development.[14]

The address created a sensation among the hard-liners of the Islamic Republic party (IRP), and the clerics clustered around Khomeini, for whom the primary purpose of the revolution was to establish an Islamic society based on the Koran. Nazih's address was regarded as—and indeed was intended to be—an attempt to influence public opinion in the struggle then under way between the moderates and the Islamic fundamentalists over the shape and direction of the revolution. Ayatollah Beheshti, a leading figure in the IRP, remarked that Nazih deserved to be tried for treason.[15]

Nazih enjoyed considerable support among both the staff and the workers at the NIOC, and the prime minister, Mehdi Bazargan, also tried to prevent his ouster. But a press campaign was organized against Nazih. "Islamic committees" were organized to complain of his mismanagement of the company and his failure to purge "counterrevolutionaries" from the NIOC staff. Khomeini's son-in-law, Ayatollah Eshraqi, appointed to investigate these charges, toured the oil fields and reported against Nazih. At the end of September the NIOC chief lost his post and, fearing arrest, fled the country.

Ali-Akbar Moinfar was appointed to succeed Nazih as NIOC chairman and managing director and also to fill the newly created post of oil

14. *Ettelaat* (Tehran), 7 Khordad 1358 (May 28, 1979).
15. *The Guardian* (London), June 4, 1979.

minister. But Moinfar, though close to some clerical circles, was not a radical. During the reign of the shah he had worked in the Plan and Budget Organization. Politically, he was associated with Bazargan's relatively moderate Freedom of Iran Movement. Upon taking over at the NIOC, he reaffirmed the government's intention to maintain production at between 3.0 and 3.5 mmbd, a figure somewhat lower than the prevailing level, but not drastically so. In fact, Moinfar went out of his way to deny reports of a drop in Iranian production. When the new budget, which Moinfar helped write, was announced in May 1980, it was found to provide for revenues from oil exports of 1,600 billion rials ($22.4 billion) for the year ending March 20, 1981.[16] This suggested an oil production target of close to 3 mmbd and a broad continuation of former policies.

The manner of Nazih's dismissal, however, had further politicized the oil industry. His removal was followed in November by the resignation of Prime Minister Bazargan, as a result of the American hostage crisis. This was a further blow to moderation and to the policy of maintaining correct relations with the United States. Even though Moinfar resisted pressures for a large-scale purge in the NIOC, retirement or expulsion of other executives and staff who had served in the NIOC under the monarchy was inevitable.

Shortly before Nazih's departure, an aide had declared the NIOC to be "more or less satisfied" with the prevailing level of official OPEC prices.[17] But after Nazih's dismissal Iran began to pursue a more aggressive pricing policy. The NIOC added premiums on top of official prices and imposed surcharges on new customers or on regular customers seeking oil purchases in excess of contracted amounts. Various formulas were devised for maximizing revenues, such as a share for Iran in the refinery margins of major customers. Stable and guaranteed supplies at a future date were made conditional on payment by customers of a higher price for current purchases. Although these measures were not out of line with the practices adopted by other OPEC members, Iran was usually the leader in initiating such arrangements and in imposing higher price increases.

Inside OPEC, on pricing and other policies, Iran tended to stand with the hawks. Moreover, Iranian delegates began to show more inflexibility. At the OPEC session in December 1979, Iran first proposed that OPEC adopt prices equivalent to the cost of producing other forms of fuel (a

16. *MEES*, vol. 23 (December 31, 1979), pp. 3–4, and vol. 23 (June 2, 1980), p. 8.
17. Interview with Movahed in ibid., vol. 22 (August 27, 1979), p. 4.

cost that an OPEC study estimated at between $35 and $60 a barrel), and then stuck to its demand for an OPEC official price of $35 a barrel long after Algeria and Libya, which had made similar proposals, abandoned this position. On the OPEC long-term strategy committee, convened to work out a formula for regular and more orderly modifications of oil prices, Iran argued for criteria that would have meant far rapider and larger price increases. When prices began to weaken, Iran argued strongly for the adoption by OPEC members of production limitations to sustain high prices.

The new militancy stemmed, in part, from the experience of the past year. Iranian crude prices had risen steadily from about $13.45 a barrel (for Iranian light) on the eve of the revolution in January 1979 to $31.00 a barrel in February 1980.[18] These steep increases were of course due to the coincidence of a number of complex market factors, as the Iranians no doubt understood. Nevertheless, it was widely believed in the NIOC and the government that an attitude of revolutionary militancy had had much to do with these price rises. The experience of conducting oil policy during the first year of the Islamic Republic thus had a powerful role in shaping the assumptions and attitudes of Iranian policymakers, and accounts in part for the inflexibility that followed.

At the same time, what might be described as the Bani-Sadr position on oil policy—that it was an absolute virtue to limit oil exports and to create, in the long run, an "oil-free" economy—gained the upper hand in government councils. Bani-Sadr himself became president in January 1980. His election reinforced sentiment at the Central Bank for holding oil revenues down to a level the country could usefully absorb. The governor of the Central Bank, Ali Reza Nobari, was a Bani-Sadr protégé. At some stage early in 1980, Moinfar also became converted to the Bani-Sadr position.

Iran's leaders could also contemplate with equanimity a sharp drop in production and exports because the country's balance-of-payments position remained strong. When the revolutionaries came to power, Iran's reserves stood at over $10 billion. By the end of the Iranian year, in March 1980—the first full year of the Islamic Republic—the balance-of-payments position on current account had improved by $4.8 billion.[19]

18. Organization of Petroleum Exporting Countries, *Annual Statistical Bulletin, 1980* (Vienna: OPEC, 1981), p. 126.

19. Interview with Ali-Reza Nobari, *Die Presse* (Vienna), November 9, 1981; and Bank Markazi Iran, *Gozaresh-e Eqtesadi-ye Sal-e, 1358* (Annual Report of the Central Bank of Iran for 1358 [1979–80]) (Tehran, 1359 [1980]), p. 92.

From prevailing oil prices (somewhat more than $30 a barrel) and projected exports of about 2.5 mmbd, it appeared to the Central Bank early in 1980 that Iran's reserves would continue to increase at a rate of about $1 billion a month.[20] The government did not want to build up such large dollar deposits in foreign banks. Besides, Moinfar was predicting oil prices of $40 a barrel by the first quarter of 1981.[21] These considerations argued for lower levels of production; and since it was expected that demand would remain firm, there also seemed to be strong arguments for pressing for high price levels.

Iran's prices were already high relative to the price of comparable gulf crudes. When in February 1980 Saudi Arabia and other gulf states raised their prices by about $2.00 a barrel, Iran pushed ahead with a $2.50 price increase. On April 1, at a time of sluggish demand and weakening prices, the NIOC announced a further price rise of over $4.00.[22] Iranian crude was then the most expensive in the gulf, between $6.00 and $8.00 above the price of other gulf crudes. This gap narrowed somewhat during the summer, as a result of production cutbacks by other gulf states, slightly firmer prices, and the elimination of some surcharges by Iran.

But Iranian prices remained high, buyer resistance developed, and Japanese companies and others refused to continue liftings at previous levels. Although Iran may have deliberately aimed at somewhat lower levels of production, it is clear from the strong denials of reports of falling liftings that the country was not prepared for the precipitous drop in exports that actually occurred. Nevertheless, Iran remained inflexible on prices. It was thus that on the eve of the war with Iraq, exports stood at about 700,000 barrels a day.

Regional Factors: The War with Iraq

The Iran-Iraq war provides an almost textbook case of the way in which regional politics impinge on oil security and supply. The war damaged refinery and oil-exporting capacity in both Iran and Iraq. In Iran it inflicted damage on installations at the oil terminal on Kharg. It interfered, at least initially, with tanker traffic on the Persian Gulf. Iraqi

20. *MEES*, vol. 23 (January 28, 1980), p. 2.
21. Ibid., vol. 23 (March 17, 1980), p. 1.
22. OPEC, *Annual Statistical Bulletin, 1980*, p. 126.

air raids and artillery fire knocked out 60 to 65 percent of Iranian refinery capacity and damaged the large and expensive joint Iran-Japan petrochemicals complex at Bandar Khomeini, then under construction. Destruction of pipelines and gas production facilities led to severe domestic shortages of bottled gas. Damage to the economy as a whole was extensive and not confined to the oil sector alone.

The war was rooted in historical frictions and conflicts between Iran and Iraq. But it would not have occurred without the Iranian revolution. The direct causes were the Iranian attempts, in a burst of ideological and revolutionary fervor, to export the revolution and the ambitions aroused in Iraqi leaders by the prospects of a quick victory over a supposedly supine Iran, weakened by eighteen months of turmoil.

Iran and the Persian Gulf States

The Iranian revolution was by its very nature calculated to cause unease among the Arab governments of the Persian Gulf. It was a revolution made in the name of Islam and the "oppressed" Moslem masses and against the "oppressors" and the powerful. It carried with it strong anti-American and anti-imperialist overtones. In the early months after the revolution, the figure of Ayatollah Khomeini seemed to have a powerful attraction for various groups in the Persian Gulf emirates. In many of these sheikhdoms, there were large Shi'ite populations and communities of merchants and workers of Iranian origin.

More significantly, Iranian leaders and clerics directly and indirectly called on their brother Moslems on the other side of the gulf to follow the example set by revolutionary Iran. Khomeini often addressed his appeals not only to Iranians but to the peoples of the Moslem world in general; his denunciation of the role of the United States, "the Great Satan," in the Moslem world, and of the Moslem governments that he believed furthered American interests, had implications for neighboring countries and for the Persian Gulf states.

After the establishment of the Islamic Republic, Iranian propagandists traveled the region carrying the message of the Iranian revolution to the gulf emirates. One of these propagandists was Hojjat ol-Eslam Sadeq Khalkhali, the judge of the revolutionary tribunals whose fame inside Iran rested on the alacrity with which he sent scores of "counterrevolutionaries" before the firing squad. Diplomats representing the Islamic Republic in the area tended to become spokesmen for a militant brand of Islam as well.

There were disturbances in at least one Persian Gulf emirate as a result of mosque sermons by Iranian representatives; Iranian diplomats were asked to leave another emirate for what was considered an undiplomatic expression of views. In Iran some clerics revived the Iranian claim to Bahrain; and the religious leader Ayatollah Mohammad Sadeq Ruhani warned the ruler of Bahrain that unless he treated the inhabitants with greater consideration, the people of Bahrain would be called on to overthrow him.

Iranian officials in responsible positions attempted to formulate a policy position that would satisfy the Islamic and revolutionary pretensions of the hard-liners without creating too much anxiety in the gulf states. The Iranian foreign minister, Ebrahim Yazdi, explained in mid-1979 that Iran had no intention of actively exporting its revolution, although it could not prevent others from drawing inspiration from the Iranian example.[23] Khomeini was persuaded to issue a statement to the same effect. When the Kuwaiti foreign minister visited Iran in July 1979, he was told that statements on the annexation of Bahrain made by various Iranian clerics or politicians did not reflect official policy.[24] Similar assurances were given to Saudi Arabia.

Kuwaiti and Saudi officials seemed to accept these assurances at face value. In public statements the Saudis went out of their way to be accommodating to the Khomeini regime. King Khaled described the establishment of an Islamic republic in Iran as a welcome development. In newspaper interviews in January 1980, Crown Prince Fahd ascribed to "irresponsible persons" sentiments hostile to Saudi Arabia expressed by various ranking Iranians. "We respect Imam Khomeini's convictions, and we will not change our attitude towards him or towards Iran," he said.[25]

Yet Khomeini continued to make militant statements about the revolution and Iranian policy toward oppressed peoples. In a message marking the first anniversary of the Iranian revolution, just one month after Fahd's conciliatory remarks, Khomeini said: "We all know that the Islamic world is waiting for the full fruition of our revolution. . . . We will export our revolution to the entire world, because our revolution

23. For the Ruhani-Yazdi exchanges, see *Bamdad* (Tehran), 25 Tir 1358 (July 16, 1979); *Bamdad*, 12 Shahrivar 1358 (September 3, 1979); and *Ettelaat* (Tehran), 27 Shahrivar 1358 (September 18, 1979).

24. *MEES*, vol. 22 (August 6, 1979), p. 3.

25. Interviews with *Al-Hawadith* (Beirut), January 11, 1980, and *Al-Safir* (Beirut), January 9, 1980, both reported in ibid., vol. 23 (January 21, 1980, supplement), pp. 5, 7.

is Islamic; until such time that the cry, 'There is no God but one God, and Mohammad is his Prophet!' does not yet ring throughout the world, the struggle continues. And wherever in the world there is struggle against the oppressor, we are there too."[26]

In practice, it also proved impossible to silence others. Ayatollah Ruhani sharply rebuked the foreign minister, Yazdi, for suggesting that Ruhani, in his remarks on Bahrain, did not speak for the Iranian government. Ayatollah Hossein Ali Montazeri, leader of the Friday prayers in Tehran and widely regarded as Khomeini's probable successor, echoed Khomeini's remarks on the need to export the revolution.[27] After his election as president in January 1980, Bani-Sadr, who was given to dramatic public statements on a wide variety of issues, was particularly vocal on the question of Iran's relations with the gulf states.

Bani-Sadr believed that the Iranian revolution would not be secure against Great Power interference unless similar revolutionary and anti-imperialist regimes were established in neighboring states. As president, he spoke contemptuously of Saudi Arabia and the gulf emirates as American "client states," which were afraid of the Iranian revolution and of their own people. "We do not consider them to be independent governments," he said, "and therefore do not wish to cooperate with them." He predicted that if the people in the Arab states opposed to Iran adopted the techniques developed by the Iranian revolutionaries, "not one of these regimes would remain in existence, and they know it. . . . All these overlords will be like dust in the wind." Iran, he said, would help other peoples to liberate themselves, not so much by material or military assistance as by the example of the Islamic government it had established. If an Islamic popular movement asked for Iran's help, he affirmed, "we would definitely help such a movement."[28]

Iran's Relations with Iraq

Iran's relations with Iraq in the post-revolution period evolved against this backdrop. There had already been a history of conflict over land frontiers and particularly over navigation on the Shatt-al-Arab, the 120-

---

26. *Ettelaat,* 23 Bahman 1358 (February 12, 1980).
27. *Kayhan* (Tehran), 6 Farvardin 1359 (March 26, 1980).
28. Interviews with *Al-Safir,* February 12, 1980, and *Al-Nahar Arab and International Weekly,* March 24, 1980, reported in *MEES,* vol. 23 (February 18, 1980), p. 2, and vol. 23 (March 31, 1980), pp. 6, 7.

mile waterway on which are located Iraq's main commercial and oil
terminals (around Basra) and Iran's largest commercial port (at Khorramshahr) and main oil refinery (at Abadan). Until 1975 navigation on
the waterway was regulated by an agreement concluded between the
two countries in 1937, when Britain still exercised considerable influence
in the region. Except for the waters directly surrounding Iranian ports,
the agreement gave sovereignty over the entire waterway to Iraq, a cause
for Iranian complaints in the years that followed.

In 1975, however, under an agreement concluded in Algiers between
the shah and Saddam Hussein (at that time deputy chairman of Iraq's
Revolutionary Command Council), Iran, which was then militarily the
stronger country, was able to secure Iraqi recognition of the "thalweg,"
or deep-water line, running roughly halfway down the river, as the
frontier between the two countries. In exchange, the shah agreed to
certain border rectifications in Iraq's favor. He also agreed to end his
assistance to Iraq's Kurdish rebellion, which collapsed a few weeks after
the withdrawal of Iranian support. By Saddam, at least, the Algiers
agreement was regarded as a humiliation for Iraq that, when the Iranian
revolution broke out, he would attempt to undo.

The revolution exacerbated other sources of friction. At least half the
Iraqi population is Shi'ite. Najaf and Karbala in Iraq are important
centers of Shi'ite learning and study; the most revered Shi'ite shrines
are located in these cities and each year draw tens of thousands of Iranian
pilgrims. Large Iranian communities live in and around the shrine cities,
for purposes of study and worship and also of trade. A tradition of
militancy exists among the Shi'ites in these cities, directed, during
various periods in recent history, primarily against the government in
Tehran but sometimes also against the authorities in Baghdad. It was
from Najaf, where he spent fourteen years in exile, that Khomeini
launched his campaign for the overthrow of the shah.

Shi'ite militancy in Iraq also expressed itself through a movement
that was founded in about 1968–69 under the name of al-Da'wa al-
Islamiyya ("The Call of Islam").[29] The facts about the origins and early
activities of al-Da'wa are in dispute. But it appears to have grown more
militant in the 1970s and to have become associated with Mohammad
Baqer Sadr, a religious teacher who sought to combine militant Shi'ism

29. An account of al-Da'wa and other Shi'ite movements in Iraq is given in Hanna
Batatu, "Iraq's Underground Shi'a Movements: Characteristics, Causes and Prospects,"
Middle East Journal, vol. 35 (Autumn 1981), pp. 578–94.

with socialism and political activism. Sadr's book on Islamic socialism, *Iqtisaduna,* became a standard text for many later Iranian writers who wished to portray Islam as a religion committed to social and economic justice. Bani-Sadr drew heavily on Baqer Sadr's theories in his own book on Islamic economics, *Eqtesad-e Towhidi* ("The Economics of Divine Harmony"). Al-Da'wa appears to have grown more active after the Iranian revolution, becoming a source of concern to the Iraqi authorities.

The obverse of this situation existed, though to a more limited degree, among the Arabic-speaking population in Iran's Khuzestan province, the center of the Iranian oil industry. Like ethnic minorities elsewhere in the country, Iranian Arabs in Khuzestan after the revolution began to organize themselves into cultural and political societies and to demand a measure of local autonomy. There were frequent clashes between them and the Revolutionary Guards. Arab political and cultural organizations were suppressed by the Iranian government, and the religious leader of the Khuzestan Arabs, Shaykh Shubayr al-Khaqani, was arrested, removed from Khuzestan by the authorities, and placed under house arrest, probably in Qom. Militant Arabs subsequently sabotaged oil pipelines and installations. Iran accused the government of Iraq of inciting the Iranian Arabs in Khuzestan, providing them with arms, and training them for sabotage activities.

On their side, the Iraqi authorities believed that Iranian revolutionary agitation, exercised through the agency of al-Da'wa and other groups, explained Shi'ite militancy in Iraq. Early in April 1980 an attempt was made on the life of Tariq Aziz, an important aide to President Saddam Hussein. The Iraqi government blamed al-Da'wa for the abortive assassination. Members of the organization were arrested, and Mohammad Baqer Sadr and his sister were later secretly executed. The Iraqi government also rounded up thousands of Iranians resident in Iraq and unceremoniously dumped them on the Iranian side of the border. According to the Iranian authorities, over 25,000 Iranians were eventually driven out of Iraq.

At the same time there were border clashes between the two countries. The Iraqis used artillery and antitank rockets to damage storage tanks at a pumping station between Naft-e-Shah and Kermanshah, cutting supplies to the local refinery. A few days later, seven pipelines serving the Abadan refinery were blown up and an attempt was made to sabotage a pipeline near Ahwaz. Iran, claiming that Iraq was massing troops on the border, put its own army in a state of alert. The April events also saw a marked increase in the intensity of the rhetoric the two countries directed at one another. In a broadcast on April 8, 1980, Khomeini described Saddam Hussein as "an enemy of Islam and the Moslems," and he called on the Iraqi people, tribes, and army to overthrow him and to "cut off the hand of America which has emerged from Saddam's sleeve."[30] Calls for the overthrow of Saddam and the Ba'ath government became a persistent theme of Iranian propaganda thereafter.

While Iranian revolutionary militancy was no doubt a cause for disquiet in Iraq, Saddam was also using tensions and border clashes to seek a revision of the 1975 Algiers treaty. Five months before the April crisis, in October 1979, the Iraqi ambassador to Lebanon had demanded in an interview that Iran "rectify its relations with Iraq" by agreeing to modifications of the Algiers agreement and restoration of Iraq's full rights over the Shatt-al-Arab, ceasing all "chauvinistic Iranian propaganda" in the Persian Gulf, and giving self-rule to the Arabs of Khuzestan, to the Kurds, and to the Baluchis. He also demanded that Iran withdraw from the three strategically located islands in the Persian Gulf—the two Tumbs and Abu Musa—which the shah had acquired in 1971 by negotiation and then outright seizure from the sheikhdoms that exercised a nominal sovereignty over them.[31]

Saddam renewed these demands in a cable to UN Secretary General Kurt Waldheim on April 6. Sporadic border clashes continued. On September 22, after announcing to his cabinet a unilateral abrogation of the Algiers treaty, Saddam sent his troops across the frontier to seize Iranian territory and his aircraft to bomb Iranian targets to enforce these various demands.

The Outbreak of Hostilities

Saddam Hussein's decision to make war against Iran was based on a miscalculation, both of Iraqi strength and Iranian weakness. He believed that Iran, racked by eighteen months of revolutionary turmoil, would be incapable of organizing a defense and that its army, weakened by purges and humiliated in other ways, would be incapable of resistance. He may also have believed that domestic dissatisfaction was so widespread that an Iraqi invasion would spark an uprising against the Khomeini regime.

30. *Ettelaat,* 20 Farvardin 1359 (April 9, 1980).
31. *Al-Nahar,* October 31, 1979, cited in *MEES,* vol. 22 (November 12, 1979), p. 5.

Iranian royalist generals and politicians in exile, who were receiving some Iraqi assistance and claimed to have knowledge of opinion within the country and the armed forces, may have encouraged Saddam in these views.

All this is generally well known. But it is useful, in terms of the dynamics of regional politics, also to examine Iranian attitudes to the possibility of open hostilities with Iraq. Although Iranian officials did not expect the sporadic border incidents and mutual recriminations to lead to open warfare, beginning in April high-ranking Iranian officials persisted, partly for domestic purposes, partly out of conviction, in a kind of reckless rhetoric about the struggle with Iraq. Officials repeatedly reiterated the view that Iran had decided to seek the overthrow of Saddam Hussein and the Ba'ath government.

Some officials and clerics held the view that, properly encouraged by example and speech, the Iraqi people would emulate the Iranian example and rise up against Saddam. Ayatollah Montazeri asserted, "Our Iraqi brothers have asked Imam Khomeini to lead their revolution." Bani-Sadr argued that if Iraq sent its troops against Iran, the soldiers "would turn their backs on Saddam."[32] The April outbreak of border fighting with Iraq coincided with the decision by President Carter to break U.S. diplomatic relations with Iran and impose trade sanctions. Many Iranian officials regarded these crises variously as an opportunity to regenerate revolutionary fervor, to unify the warring factions in Iran, to end strikes and unrest at places of work, or to crush domestic opposition.

Bani-Sadr remarked in April that "the threats from Iraq are not a negative thing; they are a positive thing. . . . This struggle gives us life; it fires and stirs us up. These people [the United States and Iraq] bring about precisely the conditions our people need for the continuation of the revolution."[33] The revolutionary prosecutor-general, Ali Qoddusi, used the crisis atmosphere to call for an end to disturbances at places of work and to warn opposition groups against creating unrest. "The day has arrived," he said, "for distinguishing the patriot from the traitor, the line of the Imam [Khomeini] from the line of Satan."[34]

Elements in the military regarded the possibility of war with Iraq as an opportunity for the army to reorganize and rehabilitate itself, in its own eyes and in the eyes of the nation. A group of army commanders

32. *Bamdad*, 25 Farvardin 1359 (April 14, 1980).
33. *Ettelaat*, 20 Farvardin 1359 (April 9, 1980).
34. *Bamdad*, 21 Farvardin 1359 (April 10, 1980).

told Bani-Sadr: "We will not be truly alive unless we march to confront the enemy and, facing death for Islam and Iran, we revive the army and turn it into a revolutionary army."[35]

There was a point at which such rhetoric shaded into conviction or reflected it. The belief that the revolution would be strengthened by struggle against foreign enemies undermined the willingness or ability of Iranian leaders to adopt measures that might have prevented the outbreak of hostilities with Iraq. When the war came, such rhetoric continued. Khomeini described the war as "not a misfortune but a blessing," because it rekindled the revolutionary spirit.

The Failure of Peace Efforts

The issues of war and peace also became enmeshed in domestic politics and political rivalries. Bani-Sadr tried to use the war to strengthen the position of the army, of which he was commander in chief, and to concentrate authority in the Supreme Defense Council, which he headed. His clerical rivals, on the other hand, sought to bolster the standing of the Revolutionary Guard and to credit it with the successful defense of Iranian cities, to discredit Bani-Sadr for his alleged mismanagement of the war, and to use the war to reinforce the authority of various revolutionary organizations. This internal rivalry, reinforced by revolutionary rhetoric, acted to make a settlement of the conflict more difficult.

Iran's peace terms required the withdrawal of Iraq from Iranian territory and recognition of Iranian rights on the Shatt-al-Arab as set out in the Algiers treaty. But Iran also had further aims, as repeatedly articulated by leading officials and clerics. These included a demand for reparations, recognition of Iraq as the aggressor, and even the punishment and overthrow of Saddam Hussein. "There can be no meaning to peace between Islam and infidels," Khomeini told a delegation from the Islamic Conference, one of three or four organizations seeking to mediate the conflict.[36]

Such rhetoric hampered peacemaking efforts. The fate of the peace proposals made public by the Islamic Conference on March 5, 1981, serves as an example. The plan called for a cease-fire, the withdrawal of Iraqi troops from Iranian territory, internationally supervised navigation

---

35. *Ettelaat*, 20 Farvardin 1359 (April 9, 1980).
36. *Financial Times* (London), March 2, 1981.

on the Shatt-al-Arab until a final peace agreement, and the settlement of all outstanding issues through negotiation or mediation. President Bani-Sadr termed the peace plan a basis for further discussion. But Ayatollah Montazeri, in a message to the Supreme Defense Council, said Iran could accept nothing less than "the punishment of the aggressor, Saddam, by an international court and liberation of our brotherly Iraqi nation from this usurper." The next day the Supreme Defense Council turned down the peace plan and reiterated Iranian demands for the punishment of Saddam.[37] Peace proposals suggested by the mediator appointed by the UN Secretary General met the same fate. Needless to say, similarly inflated rhetoric, exaggerated war aims, and inflexibility, reinforced by Saddam Hussein's domestic problems, characterized the Iraqi negotiating position.

Several conclusions can be drawn from the circumstances surrounding the outbreak of hostilities and the prosecution of the Iran-Iraq war. Saddam's national and regional ambitions, coupled with a perception of Iranian weakness, led to Iraqi aggression. Iranian revolutionary rhetoric, appeals for the overthrow of Saddam, and attempts to export revolution, though not decisive factors, nevertheless were a cause of concern to Iraq, just as attempts to export revolution became at a later date a cause of concern to Saudi Arabia and the Persian Gulf emirates. Peacemaking continued to be frustrated not only by substantive issues but also, on both sides, by domestic considerations. Even though Iran and Iraq appeared to have reached an unspoken understanding after the initial weeks of fighting to refrain from further attacks on oil and port facilities, the actual destruction inflicted on oil facilities and the damage done to oil export capacity were considerable and demonstrated the destructive potential of other such regional conflicts.

Finally, though hostilities did not spread to neighboring countries, it proved impossible to shield the region from the repercussions of the war. The war deepened cleavages in the Arab world, pushed Arab states into taking sides in the conflict, and influenced the oil policies of the gulf governments. Saudi Arabia and Kuwait placed their port and overland transit facilities at Iraq's disposal. Iran reacted by "accidentally" bombing targets in Kuwait on three occasions. In October 1981 a Kuwaiti oil and gas separation plant was seriously damaged as a result of aerial bombing. The warning to the gulf states was clear. Saudi Arabia, Kuwait,

37. For these exchanges see *Kayhan,* 11 Esfand 1359 (March 2, 1981), and 16 Esfand 1359 (March 7, 1981), p. 13.

and other gulf states extended between $20 billion and $30 billion in loans to Iraq during the first two years of the war. As the war continued, Saudi Arabia and Kuwait faced the possibility of having to adjust oil production levels to meet these (and other) substantial claims on their financial resources.

Iran and Syria meanwhile drew closer together and made common cause in seeking the fall of Saddam Hussein. Syria supplied Iran with arms, and the two countries joined hands to deny Iraq access to oil markets. Under an agreement signed in March 1982, Iran was to supply Syria with 180,000 barrels of oil a day in exchange for Syrian goods. Of the total, about 70,000 barrels were to be processed in Syrian refineries for reexport to Iran. The remaining 110,000 barrels were earmarked to meet Syria's own domestic requirements. With its crude oil needs satisfied, Syria in April closed the pipeline through which Iraq transported some 400,000 to 500,000 barrels a day across Syria to Mediterranean ports. Iraq's already meager oil exports were thus greatly diminished. Iran was reported to have supplied Syria with some crude at reduced prices to compensate Damascus for the $30 million to $40 million in transit fees it lost by shutting down the Iraqi pipeline.

Further Threats to Stability

Toward the end of 1981 and in the first half of 1982, two developments appeared further to threaten stability and oil security in the Persian Gulf.

First, Iran stepped up its political and propaganda activities in the gulf states. In the fall of 1981, for example, Iranians making the annual *hajj* pilgrimage to Mecca tried to organize demonstrations in favor of Khomeini and the Iranian revolution and there were clashes with Saudi police. That these demonstrations were perhaps not entirely a spontaneous occurrence was indicated in July 1982, on the eve of the new *hajj* season, when Khomeini appointed a radical religious leader, Hojjat ol-Eslam Mohammad Musavi-Khoeniha, as director of the Iranian *hajj* organization. Musavi-Khoeniha had been a mentor to the militant students who seized the American embassy in Tehran in November 1979. Khomeini specifically charged Khoeniha with spreading the message of the political significance of the *hajj*, which he said had been neglected.

More seriously, there were indications that Iran was prepared to finance subversion in the gulf sheikhdoms. In December 1981 the government of Bahrain arrested seventy-three persons for plotting to overthrow the government. The well-armed plotters were said to have been trained and motivated by Iran, although the Iranian government denied any responsibility. The Saudi interior minister alleged that plans to assassinate leaders in Saudi Arabia, Bahrain, and other gulf states had been uncovered. He accused Iran of training, arming, and financing a network of terrorists to undermine stability throughout the Persian Gulf, and said that Saudi Arabia was prepared to despatch troops to any threatened gulf state.[38]

Saudi Arabia later in December concluded an agreement with Bahrain and two other gulf states for cooperation on internal security matters. The Saudis made more explicit their support for Iraq in the war with Iran. The propaganda battle between Iran and the gulf states grew considerably more bitter.

Second, the tide of the war turned dramatically in Iran's favor. In three decisive campaigns—in September 1981 and in March and May 1982—Iranian forces were able to break the siege of Abadan, to push Iraqi troops out of Iranian territory along a broad front, and to recapture the port of Khorramshahr. In the face of Iranian refusal to negotiate, Iraq on June 29, 1982, announced it had completed a unilateral withdrawal from Iranian territory. Iran, however, claimed that this withdrawal was incomplete. It continued to press its demands for reparations and for the right of return for Iranians expelled from Iraq and to insist on the trial and "punishment" of Saddam Hussein. On July 13 Iranian forces invaded Iraq in an attempt to enforce these objectives.

Iran now appeared to be seeking not only the overthrow of Saddam Hussein and the Ba'ath party, but also the installation of an Islamic government at Baghdad. Moreover, threats by Iranian officials and by Khomeini himself against gulf states that were assisting Iraq grew more direct and more menacing. By the early fall, the war had ground to another stalemate, this time on Iraqi territory. But this did not rule out the possibility of an Iranian military breakthrough or attempts by Tehran, with or without decisive Iranian victory, to pressure the gulf states to adopt foreign policies and oil policies more to Iran's liking.

38. *MEES*, vol. 25 (December 28, 1981).

Economic Factors

Because of the strength of the country's foreign exchange reserves, as already noted, the Iranian government in 1980 was in a position to continue insisting on high prices for Iranian oil despite weakening world demand and also in a position to countenance the resulting steep decline in oil output and earnings that occurred in the second and third quarters of the year. Three factors, however, combined in 1981 to erode Iran's foreign exchange position: the settlement of the hostage crisis, the war with Iraq, and the growing oil glut on the world market.

The Worsening Situation

In January 1981 the Iranian government finally released the American hostages it had been holding since November 1979. The American government in turn unblocked $11 billion to $12 billion of Iranian assets held by banks in the United States and U.S. banks abroad that had been frozen by presidential order on November 14, 1979. However, of the total assets that became available to Iran, $3.7 billion was used to repay syndicated loans extended to Iranian entities by consortia of American and foreign banks. Another $1.4 billion was set aside to cover the repayment of nonsyndicated U.S. bank loans. And another $1 billion was placed in an escrow account to secure claims against Iran by nonfinancial American entities and citizens. The Iranian government thus gave up, at one stroke, over $6 billion of its foreign exchange assets.[39]

The war with Iraq, in addition to the damage it caused to ports, refineries, and petrochemical and industrial plants and to towns and cities, imposed heavy financial burdens on the country. With the Abadan refinery destroyed, additional volumes of oil products had to be imported; and the capture of the port of Khorramshahr by the Iraqis meant shipments of goods to Iran had to be rerouted to other ports at higher cost. The war created 1.8 million refugees. The cost of prosecuting the war was considerable. Although reliable data about war-imposed expenditures are not available, the figures cited by officials provide some

39. An account of the financial transactions involved in the release of the American hostages is given in Robert Carswell, "Economic Sanctions and the Iran Experience," *Foreign Affairs*, vol. 60 (Winter 1981–82), pp. 254–56.

indication. The government, for example, estimated the cost of war matériel, import of refined products, and refugee relief at $2.6 billion for the first six months of the war alone.[40] The budget for 1981–82 allocated 330 billion rials ($4.2 billion) over and above the regular defense budget to cover the prosecution of the war, war damage repair, oil-product imports, and refugee relief.[41]

The war initially disrupted crude oil exports. Liftings from the Kharg terminal were, however, resumed within weeks of the outbreak of hostilities. But discounts had to be granted to customers, in part to defray the cost of higher insurance premiums on tankers entering the war zone. On September 30, 1981, Iraqi bombers damaged the pumping station at Gurreh, the mainland facility through which crude is moved from the major fields to the terminal on Kharg. The war continued to act as a deterrent to large-scale liftings from Iranian terminals, and war damage was reported to have limited Iran's potential export capacity.

These difficulties were exacerbated, particularly in the second half of 1981, by the growing oversupply of oil in world markets and a weakening of prices. Iranian crude prices ($37 for light as of early January) appeared increasingly unrealistic under these conditions, and Iran had trouble in marketing its oil. In June the Royal Dutch/Shell Group stretched out a supply contract for 110,000 barrels a day from nine to fifteen months. A consortium of Japanese companies, with contracts to lift 220,000 barrels a day, suspended liftings for a time and subsequently elected to stretch out nine-to-twelve-month contracts to twelve-to-fifteen months. Iranian exports in the first half of 1981 did not exceed 1.0 to 1.1 million barrels a day; in the second half of the year exports hovered at between 500,000 and 700,000 barrels a day. Foreign exchange earnings at these levels were well below foreign exchange expenditures, causing the government to draw on reserves.

Iran's difficulties in marketing its oil must be viewed in the context of the generally deteriorating economic conditions in the post-revolution period. Figures published by the Central Bank and cited by officials do not yet provide a complete picture, but they give a clear indication of general trends.[42]

40. Behzad Nabavi, quoted in the *Financial Times*, April 4, 1981.
41. Mohammad Taqi Banki, in *Kayhan*, 31 Ordibehesht 1360 (May 21, 1981).
42. Bank Markazi Iran, annual reports for the years 1358 (1979–80) and 1359 (1980–81). See also economic report by Bani-Sadr, cited in the *Financial Times*, March 30, 1981, and the *New York Times*, April 6, 1981.

First, because of disruptions created by the nationalization of industry and agricultural land, the purge or exodus of managerial and professional staff, periodic shortages of raw materials and spare parts in factories and of pesticides and fertilizer on farms, and general political unrest, production fell significantly in both industry and agriculture. Most factories were reported to be operating at 60 percent of capacity or less. Many of the country's most important industrial enterprises, which had been nationalized or brought under state management, were losing money. A government spokesman, Behzad Nabavi, reported that state-operated industries in 1980–81 had lost 160 billion rials (about $2 billion), and he described this figure as an improvement over that of the previous year.[43]

Agricultural production declined for all the major crops: wheat, sugar beets, cotton, and vegetable oil. Imports of wheat in 1980–81 were 30 percent higher than in the previous year and nearly twice as high as pre-revolution levels. Cotton production in 1981 was one-third the 1978 level. Meat production fell, partly because of a failure to import sufficient feed.[44] Widespread land disputes, caused by the government's uncertain policies toward land reform, also adversely affected agriculture.

Second, the erosion of business confidence virtually killed private sector investment. Government development spending substantially slowed down as a result of the breakdown in planning and administrative mechanisms. In 1980–81 the Plan and Budget Organization was able to use only 50 to 60 percent of development allocations. Moreover, it was generally agreed that much development spending was wasteful and ineffective. The loosening of controls over the allocation and expenditure of development funds in 1981–82 intensified this process.

Substantial investments were locked up in industrial and development projects that remained incomplete. Three years after the revolution, work on the vast copper works at Sarcheshmeh in eastern Iran, reportedly 85 percent complete when the revolution occurred, was still stalled. Nor did work on the multibillion-dollar Iran-Japan petrochemical complex in southern Iran, 85 percent complete, and on the Ahway steel mill, 65 percent complete, move ahead.[45]

43. Nabavi, quoted in the *Financial Times,* April 4, 1981.

44. Interview with the director of the Cereals Organization, M. J. Asemipur, reproduced in the *Iran Times* (Washington, D.C.), 18 Ordibehesht 1360 (May 8, 1981); interview with the minister of agriculture, Mohammad Salamati, in *Enqelab-e Eslami* (Tehran), 9 Ordibehesht 1360 (April 29, 1981); and speech by Prime Minister Raja'i quoted in *Iran Times,* 1 Khordad 1360 (May 22, 1981).

45. Interview with an "industrial expert," *Enqelab-e Eslami,* 19 Ordibehesht 1360 (May 9, 1981).

Third, the post-revolution period witnessed a rapid increase in current government expenditures. This was fueled in part by the expansion of the bureaucracy in the mushrooming revolutionary organizations, such as the Revolutionary Guards, the Foundation for the Oppressed, the Crusade of Reconstruction, and the like, and in part by large government allocations to sustain failing industries, subsidize food imports, prevent unemployment, and provide various services. For example, 2,500 workers and staff remained on the Sarcheshmeh copper works payroll, and 12,000 workers on the payroll of the Ahwaz steel mill, even though both projects were idle.[46]

Fourth, because of falling revenues the government resorted to deficit spending to an unprecedented degree. The government debt to the Central Bank nearly doubled (from 142.4 billion to 270.9 billion rials) between February 1979 and March 1981. The budget deficit in 1980–81 alone came to 87 billion rials ($11 billion). And money in circulation was two and one-half times higher in February 1981 than in February 1979.[47]

Debate on the Budget

It was against this background that the government in April 1981 (or one month into the new calendar year) submitted to parliament an ambitious budget calling for 330 billion rials ($41.3 billion) in total expenditures for the 1981–82 year. Besides covering the costs of the war, meeting current expenditures, and paying for social services, the budget reflected the government's hopes of stimulating the flagging economy. Some 110 billion rials ($13.8 billion) of the total was slated for development expenditures. The budget itself was predicated on oil revenues of 2,412 billion rials (over $30 billion), or more than two and one-half times the oil income in the previous year.[48] There were, however, many practical objections to the budget as framed, and the debate over the budget proposal indicated that the controversial oil issue was still very much alive.

Revenue projections appeared optimistic, given current levels of oil production and realistic possibilities of increasing exports. Very substantial sums were allocated to revolutionary organizations. For exam-

46. *Iran Times,* 18 Ordibehesht 1360 (May 8, 1981); and *Financial Times,* April 15, 1981.

47. See Raja'i's budget message to Parliament, *Kayhan,* 3 Ordibehesht 1360 (April 23, 1981); and "Vaz'iyyat-e Eqtesad-e Keshvar (4)" in *Enqelab-e Eslami dar Hejrat* (Costa Mesa, California), 1 Bahman 1359 (January 21, 1981).

48. Raja'i's budget message, *Kayhan,* 3 Ordibehesht 1360 (April 23, 1981).

ple, the Reconstruction Crusade was given a budget of $1 billion, and the budget for the Revolutionary Guards was almost 50 percent above the previous year's level. These organizations are considered inefficient and by law are not subject to the general government audit or other stringent external financial checks.

It seemed unlikely that the government would be able to spend usefully the development funds requested. War conditions in the south-west of the country meant that many projects on paper could not be carried out. On the assumption that red tape and bureaucracy had in the past prevented a rapid implementation of development projects, the government proposed to give ministers, agency executives, and departmental heads unprecedented freedom in allocating and spending funds for development and other "urgent" needs. But in the process the usual safeguards and procedures to ensure financial accountability had also been eliminated, arousing fears that wastage and corruption would spread.

These features of the budget were sharply criticized by members of parliament and opposition groups. But much of the criticism focused on government plans for oil exports and oil revenues. Ali-Akbar Moinfar, now a deputy in parliament, argued that it was the revolution and the policies of limiting production that had led to a rise in the price of crude oil from $13 to over $32 a barrel over a brief, one-year period. He described the intention to return to "high" export levels of 2.5 million barrels a day as "shameful" and a reversion to "the evil policies of the former regime." He noted that another proponent of high export volumes was Saudi Arabia's Sheikh Yamani, whose policies were designed to further the interests of his American "masters," and whose government was "totally under the influence of Western imperialism." Those who were suggesting that Iran export 2.5 million barrels of oil a day, he remarked, were pursuing similar policies.[49] Oil exports at this level were also criticized as inflationary and unnecessary by the deputy Yadollah Sahabi, a member of the cabinet of Mehdi Bazargan, whose Freedom of Iran Movement was now supporting Bani-Sadr.[50]

The sharpest attacks on the budget came from Bani-Sadr himself, who described the budget as inflationary and the removal of controls over spending as dangerous. He attacked the proposal to increase oil exports as serving the interests of the "world exploiters"—the industrial

49. *Enqelab-e Eslami*, 9 Ordibehesht 1360 (April 29, 1981).
50. Ibid., 24 Ordibehesht 1360 (May 14, 1981).

states—which would secure cheap oil, and of the reactionary Arab states, which would be able to conserve their oil while Iran squandered its mineral resources. "This action," he stated, "poses a real danger to the independence of the country. Such a step intensifies the vicious circle of poverty and dependence" and "sucks our blood like a leach."[51]

Bani-Sadr's criticisms were designed in part to cause difficulty for the government. Although he was still president, Bani-Sadr was already describing himself as "the leader of the opposition." But his criticisms were also rooted in conviction. Bani-Sadr considered the reduction of Iranian oil exports as one of the great achievements of the revolution. His views on ending what he regarded as foreign exploitation of Iran's resources and the country's dependence on oil were well known. He was supported in his stand by economists in the Plan Organization and the Central Bank who, though not sharing all his views, sought as they did under the shah to control wasteful government spending by restricting access to revenue. One group in the Plan Organization had, in fact, submitted proposals for a budget based on oil exports of about 1.3 mmbd.

The public debate on oil exports and the uses to which oil revenues were being put had embarrassing consequences for the government. Bani-Sadr's views echoed sentiments shared by left-wing political groups and also spoke to a belief among the middle class that Iran's oil resources and revenues were going to waste. Criticism of government policy was sufficiently strong to make the government issue a long refutation of Bani-Sadr's allegations. The reply, written by the prime minister's deputy for budget affairs, Mohammad Taqi Banki, appeared in successive issues of the large Tehran papers.[52] The public debate also made an impression on the parliament.

In committee and on the floor of the house, the legislature forced the government to reduce its oil revenue projections from $30.0 billion to $18.7 billion (predicated on crude exports of 1.5 mmbd) and to cut the defense budget by one-third, the development budget by about one-quarter, and the budget for the government's current expenditure by 5 percent. Despite these cuts in expenditure, lower anticipated oil revenues left the government with a deficit of 70.2 billion rials ($8.5 billion), which it said it would cover with an unspecified mix of higher tax revenues, borrowing, and increased oil exports "if the price is right."

51. Ibid., 16 Ordibehesht 1360 (May 6, 1981).
52. Banki's reply appears in a series of articles published in *Kayhan* between 31 Ordibehesht and 10 Khordad 1360 (May 21–31, 1981).

In fact, as noted earlier, the government was unable to achieve even these more modest crude export targets. Exports in the Iranian year ending March 20, 1982, averaged under 1.0 mmbd, as against a projected goal of 1.5 mmbd. Moreover, prices in the latter part of the year were lower; some crude, lifted under barter arrangements, did not earn hard currency. By the fall of 1981, serious foreign exchange difficulties necessitated a severe curtailment of imports and other austerity measures.

For nearly three years Iran had made oil policy a hostage to ideological considerations and internal political rivalries. Officials had again and again misjudged the state of the oil market. The revolutionary government had decimated the ranks of its own managerial staff in the oil industry and mismanaged the economy. It had stumbled into an unwanted war. As a result, the Islamic Republic in early 1982 found its foreign exchange reserves nearly exhausted, its traditional marketing networks disrupted, and buyers for its oil hard to find.

A Revival of Pragmatism

These problems account for indications in late 1981 and early 1982 of a more pragmatic approach to oil policy. In the fall of 1981 officials began to speak of the need to increase exports and to recapture traditional Iranian markets. Reflecting the new official position, an editorial in *Kayhan* on November 22 argued that while ending dependence on oil must remain a long-term goal, in the short run oil revenues would have to finance the reconstruction of the economy and the confrontation with domestic and foreign enemies. Two developments facilitated the return to pragmatism. First, the impeachment of President Bani-Sadr in June 1981, and the suppression of his followers, other opposition groups, and the independent press in the months that followed, eliminated potential critics of a more flexible oil policy. Second, at the OPEC conference in Geneva in October 1981, members agreed to reduce the differential affecting very light grades of crude and set a new marker price of $34 a barrel for Arabian light. These decisions demonstrated that unrealistically high prices could not be maintained in a soft market and provided a face-saving formula by which the Iranian authorities could reduce Iran's own inflated prices.

Iran continued to encounter difficulties in disposing of its oil. It was attempting to reenter the market at an unfavorable time. Other producers

(Mexico, Great Britain) were cutting prices; other OPEC states, hard-pressed for funds, were seeking to increase crude exports. Moreover, what with Iran's reputation for unreliability as a supplier, the risks that tankers faced in entering the war zone, and the Iranian refusal to sell to American oil firms, buyer resistance remained strong. Also, Saudi Arabia may have been keeping production high in a deliberate attempt to deprive Iran of a market for oil. This was mentioned by the Saudis as a way to pressure Iran to end the war with Iraq and to cease its subversive activities in the gulf.

Iran, however, pursued a more aggressive marketing policy. It cut prices by $1 on February 5, 1982, by another $1 on February 12, and by $2 more on February 21. It offered larger discounts under the table and on the spot market; it entered into barter agreements to exchange oil for food and other goods; and it made continued access to Iranian markets for Iran's major suppliers contingent on purchase of Iranian oil. These measures gradually overcame buyer resistance. Moreover, Iran refused to adhere to the production-quota plan that OPEC had adopted in March 1982 to try to prevent a further decline in prices. The OPEC states agreed to limit total production among members to 17.5 million barrels a day, setting production limits for each OPEC country. Iran's quota was set at 1.2 mmbd. But by the end of June Iran was producing an estimated 2.2 mmbd, and perhaps slightly more. At the OPEC meeting in July, the Iranian delegation argued that Saudi Arabia should cut back its production (then running at around 6.5 mmbd) to 5.0 mmbd, thus permitting Iran to raise its production to 2.5–3.0 mmbd. Iran rested its claim to a larger share of the market on size of population, need, and "historical share"; and, interestingly, the Iranian position was supported by several other OPEC states eager to see Saudi Arabia reduce its production so that they could increase their own. To many observers the July conference seemed to mark the re-emergence of Iran as an assertive and influential member of OPEC.

Conclusions

Several conclusions can be drawn from a study of Iranian oil policy during the first three years of the Islamic Republic:

—Given the assumptions of the revolutionaries who came to power, Iranian policies regarding pricing, production, and dealings with the international majors were bound to change following the revolution. But after a brief, initial period of relative moderation, more extreme lines of policy were adopted, because of ideological considerations, internal factional struggles, capture of the government by more radical elements, and a misreading of the international oil market.

—The reversion to more pragmatic oil policies was dictated by the decline in foreign exchange reserves and in general economic conditions. However, in the prevailing atmosphere, officials were not always free to follow their inclinations regarding pricing and production. The attempt of the Raja'i government to increase production to 2.5 million barrels a day under the 1981–82 budget, for example, was frustrated by political rivals who created an outcry over the prime minister's alleged intention to market Iranian oil at giveaway prices. By the spring of 1982, when the government was facing severe foreign exchange shortages, the opposition had been silenced and the government was able to reduce prices and raise production. The issue, however, remains a potentially troublesome one for the government.

—By cutting prices and raising production in the spring and summer of 1982, Iran displayed a willingness to flaunt OPEC decisions to satisfy its own revenue requirements. But Iranian authorities also indicated some appreciation for the value of OPEC as an instrument through which the gulf producers could coordinate policy and through which Iran could secure its own ends. Iran, for example, sought support for its policies among members and tried to challenge Saudi Arabia's predominant role in the organization. The Islamic Republic can be expected, in the future, to be at once more assertive in OPEC and to go its own way in determining oil policy if pressing domestic economic needs so dictate.

—By the summer of 1982 Iran was displaying a stronger desire to regularize relations with the major, non-American oil companies and to regain its traditional markets. Marketing patterns for Iranian oil, however, remained mixed, with large shares of total exports being sold on the spot market, through barter agreements with East bloc states and countries like Turkey and Pakistan, and to nontraditional buyers like Spain. This marketing pattern seemed likely to continue, or to change only slowly.

—The continuing Iran-Iraq war constituted an element of uncertainty whose effect on oil production in both countries remains difficult to measure. On the one hand, the cost of the war reinforced the inclination in both Tehran and Baghdad to raise oil production as a means of increasing revenues. For Iraq, where a production increase was not immediately feasible, this implied a continuing reliance on Saudi Arabia and Kuwait as a source of funds. On the other hand, the war could at any moment further disrupt the oil exporting capacity of Iran or Iraq.

—In the near future Iranian production levels appear likely to be determined primarily by immediate economic requirements. It is generally estimated that Iran requires a minimum of $13.0 billion to $15.5 billion in oil revenues a year to finance its basic requirements for imports of food, raw materials, spare parts, equipment, and essential consumer goods. Depending on the price of oil—say between $25 and $30 a barrel—this would argue for minimum exports of between 1.2 and 1.5 million barrels a day, with a further 500,000 barrels to meet domestic requirements. However, Iran's potential revenue needs are much greater. The war with Iraq is costly. The government faces a pressing need to repair the war-shattered economy, to meet public demands for housing and social services, and to resume work on large industrial and infrastructure projects. These programs entail a high foreign exchange component.

BY THE SUMMER of 1982 Iranian officials were themselves speaking of a production target of 2.5 to 3.0 mmbd, or exports of between 2.0 and 2.5 mmbd. By resorting to discounts, spot market sales, and barter arrangements, officials had managed by July 1982 to raise production to around 2.2 mmbd. Because of their somewhat makeshift nature, and because barter agreements do not yield foreign exchange, these arrangements are not entirely satisfactory to Iran. In seeking to increase production and reestablish secure markets, the government was thus facing three important tasks.

First, it needed to repair its relations with its major, traditional customers and to reestablish a reputation for reliability and consistency as an oil supplier. This could be achieved only over the longer term. Second, it needed to reach some understanding on prices, production levels, and market shares with its OPEC partners. Such an understanding did not appear out of reach but was hampered by continuing friction between Iran and Saudi Arabia and to a degree by the war with Iraq. Third, despite Iranian success in raising production while still continuing the war, Iran needed to bring the conflict with Iraq to an end.

The successful pursuit of these goals presumed an ability to plan and coordinate long-term energy and foreign policy goals and to temper ideological fervor and internal political considerations to these aims. By mid-1982 there were signs of a returning pragmatism in the area of oil policy. But the conditions that had contributed to volatile and erratic oil policies in the recent past had not been eliminated.